Understanding the Food Chain

Orlando Austin New York San Diego Toronto London

Visit *The Learning Site!*
www.harcourtschool.com

Food for All

Every living thing needs food. Both plants and animals need food to grow and stay healthy. But different organisms get food in different ways.

Plants make their own food by using sunlight, air, and water. Plants are producers. **Producers** are living things that make, or produce, their own food. The food is full of the energy they need. Grasses, bushes, and trees are producers.

Apple trees are producers. They make their own food from sunlight, air, and water.

You might eat an apple for food, but you did not "make" the apple. Even the farmer who grew the apple did not "make" it. The farmer gathered it from a tree that produced the apple.

Animals can't produce their own food. Animals are consumers. **Consumers** are living things that get energy by eating other things, either plants or the animals that eat the plants.

Some consumers, such as birds, eat plants. Energy from the plants is taken into the birds' bodies. Other consumers get energy by eating the birds.

A **decomposer** is a living thing that breaks down dead things for food. Earthworms and mushrooms are decomposers. Other decomposers are so small that you can see them only with a microscope.

Focus Skill **COMPARE AND CONTRAST** **What is the difference between producers and consumers?**

This salad is full of producers. It will be eaten by a consumer, maybe by you!

What's on the Menu?

There are three kinds of consumers. Each kind eats from a different menu.

Herbivores eat only plants. A garden snail is a small herbivore, weighing only about 50 grams (1.75 oz). A dromedary camel is a large herbivore, weighing about 400–600 kilograms (880–1,326 lb).

Each herbivore has special body parts that help it eat plants. A garden snail has a tongue covered with tiny, curved teeth. The teeth help tear off pieces of leaves to eat. The dromedary camel lives mostly on thorny plants. It has lips that are thick and tough.

This herbivore eats the nuts and berries from plants.

A **carnivore** gets its food by eating other animals. Carnivores have body parts to help hunt and eat. Most frogs are carnivores that have long, sticky tongues to help catch insects. Lions have sharp teeth to help catch and eat the animals they hunt.

An **omnivore** eats both plants and animals. Freshwater mussels live at the bottoms of rivers and lakes. They eat microscopic particles—both animals and plants—out of the water.

Most omnivores have teeth that help them eat both plants and animals. Sharp teeth in the front of a human's mouth help tear meat. Flat teeth in the back help grind plants.

COMPARE AND CONTRAST How are carnivores and herbivores alike? How are they different?

A house cat is a carnivore.

Predators and Prey

You know that carnivores and omnivores eat other animals. Some need to hunt for their food. A **predator** is an animal that hunts another animal for its food.

Lions are predators. They can sneak up on other animals and overpower them. Grizzly bears are powerful predators, too. Not all predators are as large as lions or bears. Anteaters are predators. They hunt ants by using their tongues. Spiders are predators. They set traps in their webs and wait for a meal to come to them.

An animal that is hunted by a predator is called **prey**.

A wild giant anteater consumes an amazing number of insects—sometimes up to 30,000 in just one day!

Fast Fact

Pollution can hurt sea turtles. Plastic bags and balloons look a lot like jellyfish to sea turtles. When turtles eat the bags, the turtles get very sick.

Jellyfish are prey for sea turtles.

Zebras are prey for lions. Fish, deer, and squirrels are prey for bears. Insects are prey for anteaters and spiders.

Some animals can be both predators and prey. A tuna fish might hunt and eat a shrimp. Tuna fish are predators of shrimp. But then the tuna might be eaten by a shark. Tuna fish are prey of great white sharks.

MAIN IDEA AND DETAILS **A cat hunts for mice. Which is the prey? Why?**

Food Chains

A **food chain** shows the path of food from one living thing to another. There are many food chains. You could be at the end of many different food chains.

One food chain begins with an orange that gets its energy from sunlight. The orange tree stores the sunlight's energy in its oranges. When you eat the orange, you get the energy that was stored in the orange. This energy first came from sunlight.

Another food chain begins with corn that gets its energy from sunlight. A cow eats the corn. Then the cow is milked. You drink the milk. The energy you get from the milk started with the energy from sunlight.

In this food chain, a rabbit eats grass. Later, a larger animal may eat the rabbit.

The food chain can have predators and prey. An insect is the prey of this brown bat.

Some food chains are long, and some are short. A food chain always begins with a producer. In food chains that contain animals, the smaller animals are usually eaten by the larger animals. The smaller animals are prey. The larger animals are predators.

When an animal or a plant dies, decomposers become part of the food chain. For example, cows might not eat all the corn that has been brought to them. The corn that falls to the ground might be broken down by decomposers. It then becomes part of the soil again.

SEQUENCE What happens to the energy in a food chain when a squirrel eats an acorn?

Food Webs

Most animals eat many different things. A tuna might eat a shrimp or a crab. You might eat the tuna. A chicken might eat corn or an insect. You might then eat a chicken's egg. A banana might grow in the forest. You might eat the banana.

All of these are food chains. You and most animals are part of many different food chains. Many food chains overlap. Overlapping food chains are called **food webs**.

Many things can change in a food web. If the weather makes more plants grow, there could be more herbivores in a food web. If there are more herbivores, there might be more carnivores. If a wetland is drained or damaged by pollution, there might be fewer herbivores and carnivores.

Different animals—fish or turtles—might eat this algae. These algae are part of many overlapping food chains, or a food web.

In Australia, the cane toad is changing food webs.

If a new plant or animal is added, a food web can change. A new plant might crowd out other plants. Or a new animal might eat many of the plants and other animals.

Cane beetles once destroyed sugar cane crops in Australia. So cane toads were brought there to eat the beetles. But the cane toads also ate many of the other animals in the area. The toads have now crowded out some kinds of animals.

Fast Fact

The many species of algae on Earth capture more of the sun's energy and produce more oxygen than all plants combined.

MAIN IDEA AND DETAILS **What might happen to a food web if there are fewer plants?**

Energy Pyramids

Food is what animals use to get energy.

Producers use the sun's energy to make food.

A consumer, such as a rabbit, takes in the energy when it eats the plant. The rabbit gets energy from the plant.

A wolf might then eat the rabbit. The wolf gets energy from the rabbit.

In this simplified food chain, there are predators and prey. There are producers and consumers. There are herbivores and carnivores. But sunlight is the source of all the energy passed along the chain.

An **energy pyramid** is a diagram that shows how energy gets used in a food chain.

The biggest part of the pyramid, on the bottom, is for the producers. The energy is passed from them to others in the pyramid.

Herbivores come after producers. They use some of the energy. The rest of the energy passes to the animals that eat the herbivores.

SEQUENCE Explain how energy moves from producers to carnivores.

AN ENERGY PYRAMID
Carnivores get energy from the herbivores.
Herbivores get energy from the producers.
Grass and plants are the producers.

Staying Alive

Different animals find different ways to defend themselves and stay alive within their food chain.

Small insects may seem to be easy prey for the predators that want to eat them. But some have colors that blend into their surroundings. This makes them hard to find. Other insects give off bad odors. Some have hard outer skins that make them difficult to eat.

A rabbit might seem easy to hunt by a larger animal such as a wolf. Rabbits have good hearing and can tell when predators are coming. They can run very fast and turn quickly. This helps them get away from danger.

MAIN IDEA AND DETAILS **Why is sunlight important to a food chain?**

A ladybug gives off a bad odor that keeps predators away.

Energy can be tasty! Most people get their energy from plants and animals.

Summary

All living things need energy. Producers make their own food. Consumers must eat plants or animals. Decomposers get energy from dead matter. Herbivores eat only plants. Carnivores eat only animals. Omnivores eat both plants and animals. A predator hunts other animals, its prey, for food. In a food chain, producers give energy to animals that eat them. Most food chains overlap and form food webs. An energy pyramid shows how energy moves through a food chain. Sunlight is the first source of energy.

Fast Fact

Camels have adapted to survive in the desert. They have three eyelids to protect themselves from blowing sand.

Glossary

carnivore (KAHR•nuh•vawr) An animal that eats other animals (5, 6, 10, 12, 13, 15)

consumer (kuhn•SOOM•er) A living thing that gets its energy by eating other living things as food (3, 4, 6, 12, 15)

decomposer (dee•kuhn•POHZ•er) A living thing that breaks down dead organisms for food (3, 9, 15)

energy pyramid (EN•er•jee PIR•uh•mid) A diagram that shows how energy gets used in a food chain (12, 13, 15)

food chain (FOOD CHAYN) The path of food from one living thing to another (8, 9, 10, 12, 14, 15)

food web (FOOD WEB) Food chains that overlap (10, 11, 15)

herbivore (HER•buh•vawr) An animal that eats only plants (4, 5, 10, 12, 13, 15)

omnivore (AHM•nih•vawr) A consumer that eats both plants and animals (5, 6, 15)

predator (PRED•uh•ter) An animal that hunts another animal for food (6, 7, 9, 12, 14, 15)

prey (PRAY) An animal that is hunted by another animal, a predator (6, 7, 9, 12, 14, 15)

producer (pruh•DOOS•er) A living thing that makes its own food (2, 3, 9, 12, 13, 15)